大型纪实性史诗木偶剧
EPIC DOCUMENTARY DRAMA PUPPET SHOW

大象来了

HERE COME
THE ELEPHANTS

中国野生动物保护协会
中国木偶艺术剧院◎编著
BY CHINA WILDLIFE CONSERVATION ASSOCIATION
AND CHINA PUPPET ART THEATRE

中国海关出版社
China Customs Press
·北京·

2021 年春天，15 只大象组成的"短鼻家族"从云南西双版纳一路向北迁移……

这是一群可爱的"**中国故事讲述者**"，我们一起帮助他们重返家园吧！

In the spring of 2021, the "Short-trunk Family"—a herd of fifteen elephants—set off their northbound journey from Xishuangbanna, Yunnan Province…

What a lovely troop that brings us a wonderful story of China. Let us help them return to their homeland.

谨以此书献给
对自然充满敬畏之心
为构建人与自然和谐共生的地球家园而努力的人们

For those who revere Nature, and strive to build a harmonious,
mutually beneficial relationship between human beings and Nature on our planet.

大型纪实性史诗木偶剧《大象来了》
—— 主创团队名单 ——

出品单位：北京演艺集团
　　　　　中国木偶艺术剧院
合作单位：央视动漫集团有限公司
　　　　　北京广播电视台卡酷少儿卫视
　　　　　北京新京报传媒有限责任公司
指导单位：中国野生动物保护协会

出 品 人：赵佳琛
联合出品人：余俊生　刘军胜　蔡志军
总 策 划：王珏
总 监 制：董宁　金黎明

制 作 人：赵永庄
艺术顾问：王林　讲武生
科学顾问：武明录
策　　划：秦新春　成惠彬　王健
监　　制：张宾　史炳文　郭艳如

技术总监：王健
舞台监督：田宇　王昕
灯光技术：王玥　郭昊鑫
视频技术：郭昊鑫
音频技术：杨洋　李扬　王兴
舞台技术：张福稳　董洪星
道　　具：韩彦荣

编　　剧：石学海
导　　演：石学海
执行导演：宋泽超　陈曦
舞美设计：陈曦
视频设计：陈曦
灯光设计：王琦
作　　词：石学海
作　　曲：杨青　石天戈
舞蹈编导：张颖
偶型设计：李予多
音频设计：宋泽超
宣传策划：俞燕

角色表：
象 妈 妈：律海波　李雅茹
象 爸 爸：李洪博　曹乾
猛　　子：王亚妮　张姝
新生小象：杨睿
兰　　草：张姊　贺已琢
奶　　泡：王昕　王方
惹　　惹：周萌　吴旭
宽　　宽：代音音　常英
群 小 象：周苗苗　张蔚
群　　象：王秋敏　张立立　支道颖　柳蕊娟　王子　孙克克　田宇
　　　　　高平　杨晨馨　梁馨月　王秋瑾　张姊　郑杨　刘雅芳
动物保护工作者：阎爽

中央广播电视总台虎年春节联欢晚会《万象回春》节目
—— 中国木偶艺术剧院演职员名单 ——

象 妈 妈：律海波　李雅茹
象 爸 爸：李洪博　曹乾
象 哥 哥：王亚妮　张姝
象 妹 妹：周萌　吴旭
新生小象：杨睿
群 小 象：周苗苗　张蔚　贺已琢
群　　象：张姊　王方　代音音　常英　梁馨月
　　　　　支道颖　高平　张立立　杨晨馨　王秋敏
　　　　　柳蕊娟　孙克克　王子　田宇　王昕

穿　　象：王秋瑾　郑杨　刘雅芳　韩彦荣
　　　　　刘璐娅　李子瑄
领　　队：俞燕　宋泽超　陈曦

目录
CONTENTS

第一幕
Scene 1

向北向北

To the North, to the North

清晨，西双版纳雨林深处，象爸爸、象妈妈带着"短鼻家族"激情地奔跑着……突然，小象兰草摔倒了，猛子急忙上前把她扶了起来。

Dawn graces the heart of Xishuangbanna. With Papa Elephant and Mama Elephant in the lead, the "Short-trunk Family" cavort across the rainforest... Suddenly, Lancao trips. Mengzi comes to her aid all at once.

"短鼻家族"开始了他们的北移之旅。他们哼着高亢低回、激情涌动的小曲，逶迤前行着。

The "Short-trunk Family" is on its migration to the north. Every step along their meandering path is accompanied by the melodious, heart-lifting tunes they hum.

《向北向北》
To the North, to the North

第二幕
Scene 2

生平第一课

The First Lesson in Life

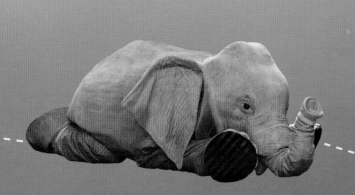

猛子突然冲了出来，要和奶泡来场较量。一场一触即发的"战争"开始了。他们的头顶到了一起。

Dashing out from the crowd, Mengzi charges at Naipao. A fight is about to break out. Their heads collide.

猛子憋足劲儿，把奶泡顶出老远。不过，猛子并没有得意多久，因为此时憨憨站了出来，要和他对决。他俩你来我往拉起锯来。

Hitting with all his strength, Mengzi sends Naipao flying. He does not bask in his victory for long, however, because Hanhan stands up to him. These two end up exchanging many well-matched blows.

13

猛子赢了。他更得意了，在象群中转来转去，彰显着自己的实力。

这时，猛子看到象爸爸严肃的表情，想要溜走，却被象爸爸用鼻子拉了回来。猛子和象爸爸开始了极其紧张的对决。

Mengzi comes out winning again. Prouder than ever, he struts about, flaunting his might—until he sees Papa Elephant's solemn brow.

Hoping to slink away, Mengzi is nonetheless pulled back by Papa Elephant's trunk. Mengzi and Papa Elephant clash, while others hold their breath.

象爸爸经验丰富，一个反转把猛子顶翻在地。突然，象妈妈一声长吼，生下了一只湿淋淋的象宝宝。象宝宝摇摇晃晃地想站起来，艰难地往前挪动，结果扑通倒在地上。

Papa Elephant, full of experience, throws Mengzi down in one tricky move. Out of the blue, Mama Elephant howls as she gives birth to Baby Elephant, who is drenched in fluids. Struggling to stand up and walk, the tottering Baby Elephant soon falls back onto the ground.

象爸爸赶忙上前要把象宝宝扶起来，却被象妈妈拦住了。象宝宝躺在地上，急促地呼吸着，尖细地哭叫着，缓慢地翻动着。

Papa Elephant hurries toward Baby Elephant to help him get back onto his feet, but Mama Elephant stops him. The feeble baby turns over on the ground, his quickened breathing interrupted by reedy cries.

象爸爸实在不忍心，又要上前，又被象妈妈拦住了。

Too soft-hearted to stand by, Papa Elephant steps forward again.
However, Mama Elephant stands her ground.

象宝宝在妈妈肚子里待了 22 个月（600 多天），终于来到了这个世界。此时他用尽全身力气，支撑着身体想站起来，但还是瘫倒在地上。

Baby Elephant has lived in Mama Elephant's womb for twenty-two months (which is more than 600 days). Having said goodbye to his former home, he uses up every last bit of energy to get up, yet meets only failure.

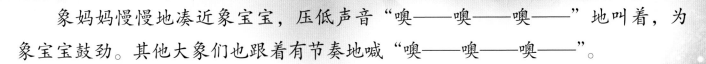

　　象妈妈慢慢地凑近象宝宝，压低声音"噢——噢——噢——"地叫着，为象宝宝鼓劲。其他大象们也跟着有节奏地喊"噢——噢——噢——"。

　　伴随着一次又一次的加油声，象宝宝仿佛得到了莫大的鼓励，他艰难地支起腿，要往前挪动……

Mama Elephant gingerly approaches Baby Elephant, encouraging him with her soft, rhythmic voice, as if beating time. *Pawoo—pawoo—pawoo—*. Other elephants follow her lead. *Pawoo—pawoo—pawoo—*.

Their waves of cheering boost Baby Elephant's courage. He does the best he can to straighten his wobbly knees, so that he may walk…

象宝宝终于摇摇晃晃站了起来，他稳了稳神，站好了，开始试探着往前迈步。一步，两步，三步……

Finally, Baby Elephant manages to stand up, although a bit unsteady. After taking a brief rest and finding his balance, he tries to throw one foot ahead of the others, one step, two steps, three steps…

象宝宝成功了！他能自己行走了！象妈妈激动地跑过去，用自己的头拼命地蹭着象宝宝的头。

"短鼻家族"一起欢呼着，围绕着象妈妈和象宝宝跳起了欢快的舞蹈。

Baby Elephant has made it! He can now walk on his own! Mama Elephant rushes to his side and nuzzles his head in great excitement.

All the members of the "Short-trunk Family" trumpet with joy. They dance around Mama Elephant and Baby Elephant to celebrate.

象宝宝开心地展示着自己，有时候得意忘形地满地打滚。象宝宝跳着跳着，突然安静了下来，在空旷的地上，躺着睡着了，鼻子里发出细微的鼾声。

Baby Elephant shows off his new ability. Growing dizzy with his success, he rolls all about on the ground. His frolic comes to an abrupt end as he falls asleep. Light snores escape his trunk as he lies on the open ground.

23

夜晚，一轮圆月从山林后面缓缓升起，象妈妈陪在象宝宝身边，用鼻子摇动着树枝，为象宝宝驱赶蚊子。

A full moon rises slowly from behind the green hills. Mama Elephant stands next to Baby Elephant, waving about a branch with her trunk—it is for chasing the mosquitoes away from her baby.

《祝愿天下母亲平平安安》

Wish Peace and Safety to All Mothers

作词：石学海
作曲：杨 青 石天戈

扫码听歌

猛子醒了，看了看象妈妈，又看了看熟睡的伙伴，悄悄地离开了象群。

Mengzi wakes up. He looks at Mama Elephant as well as his companions, and finds them all fast asleep. He then sneaks away.

放飞

Letting Go

清晨，象爸爸叫醒了"短鼻家族"的成员。他们在象妈妈的指挥下，开始了新一天的征程。

At the break of the day, Papa Elephant wakes members of the "Short-trunk Family" up. Led by Mama Elephant, the herd sets off once again.

大象们甩着鼻子行走在雨林附近的村落里。一路上，奶泡和宽宽互相嬉闹着，你踢我一脚，我再来个回旋踢。

象宝宝在泥里打滚玩耍。

The elephants pass by a village at the edge of the rainforest, their trunks swinging like a pendulum. Naipao and Kuankuan have been playing together along the way, trading all sorts of kicks.

Baby Elephant takes to the mud.

猛子的淘气之举让象妈妈心神不宁，她停下来四处张望。象妈妈向远处呼唤着，一声又一声，焦虑又急促。小象兰草陪在象妈妈身边，也跟着呼唤起来，希望猛子可以听见。

Worried about naughty Mengzi, Mama Elephant stops to look around for him. Again and again, she calls out his name in a high voice filled with anxiety. Lancao, who stands next to Mama Elephant, follows her suit, hoping Mengzi will hear them.

等到整个"短鼻家族"都走远了，猛子从藏着的地方走了出来，看着走远的象群，暗自窃喜。

猛子开始在雨林中撒欢。

Once the "Short-trunk Family" has left, young Mengzi comes out from his hiding place. He chuckles to himself as he watches his herd disappearing from his sight.

He begins to mess around in the rainforest.

《放飞自我》
Let Myself Go

作词：石学海
作曲：杨 青 石天戈

都说我长得猛　都说我长得俏　我要展示才华看看 能耐有多少

呀呀呀呀呀呀　呀呀呀呀呀呀　我撒欢地跑 撒欢地跳

都说我长得猛　都说我长得俏　我要展示才华看看 能耐有多少

呀呀呀呀呀呀　呀呀呀呀呀呀　我撒欢地跑 撒欢地跳

呀呀呀呀呀呀　呀呀呀呀呀呀　我很爽也很恐慌 自己跟自己干杯

呀呀呀呀呀呀　呀呀呀呀呀呀　喜欢就是原动力我的

目标是长大　呀呀呀呀呀呀　呀呀呀呀呀呀 酸甜 苦辣

我 愿 意　欢笑 眼泪　不 后 悔

呀 呀 呀 呀 呀 呀　呀 呀 呀 呀 呀 呀

第四幕
Scene 4

迷醉

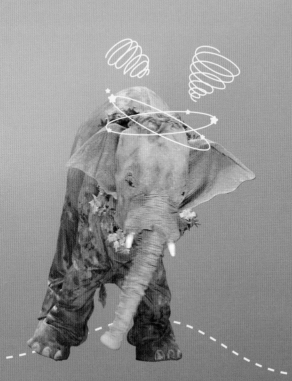

Getting Drunk

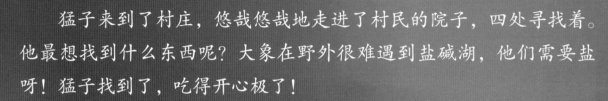

猛子来到了村庄，悠哉悠哉地走进了村民的院子，四处寻找着。他最想找到什么东西呢？大象在野外很难遇到盐碱湖，他们需要盐呀！猛子找到了，吃得开心极了！

Entering the village, Mengzi strolls into a villager's yard. What is he looking for? Salt lakes are hard to come by in the wild, but elephants need salt! Having found some, Mengzi happily stuffs his face.

猛子从屋子里走出来，用鼻子拎着饭锅，往地上一放，甩了甩鼻子，这才用鼻子往嘴里扒拉着米饭吃起来。

Mengzi comes out of the house, holding a rice cooker with his trunk. He settles it on the ground, swings his trunk in anticipation, and then digs in, using his trunk to scoop out rice and send it into his mouth.

猛子又看到院子里挂着一串辣椒，火红火红的。猛子尝了尝，被辣得在院子中央直打嗝。这时，象妈妈和象群出现在山坡上。

Mengzi finds a braid of bright red chili hanging in the yard. He takes a bite. The chili is so hot that he cannot help but hiccup in the middle of the yard.

Meanwhile, Mama Elephant and the herd appear on the hillside.

36

他们继续呼唤着猛子。

象妈妈和大家都把鼻子触到地上，发出"呜——"的长音。

在院子里的猛子听到了，却故意不理睬。

They keep calling for Mengzi.

Mama Elephant and others lay the tip of their trunk on the ground, and make a long *woooooo*.

Back in the farmer's yard, young Mengzi hears them, but acts as if he hears nothing.

猛子来到院子角落里，看到一个池子。他抬起鼻子闻了闻，一下把鼻子伸进池子里，拼命地喝了起来。

这个小伙子，没找到水，却找到了酿酒用的酒池子，这下好了，他咕咚咕咚一口气喝完……像一个醉汉在院子里打起了醉拳。

In the corner of the yard, Mengzi discovers a reservoir. He gives it a sniff, and then buries his trunk inside, drinking to his heart's content.

Instead of water, the boy has found a reservoir of wine. Having gulped down the wine in one go, Mengzi turns into a drunken kung fu master.

猛子终于醉倒在地上。

At last, Mengzi passes out on the ground.

过了一会儿，猛子又站了起来，在院子里翻腾着，一不小心钻进了栅栏里，鼻子被夹住了。他猛一使劲，把栅栏拔了出来，鼻子却仍牢牢卡在栅栏上。猛子甩了半天才甩掉。

　　猛子急忙跑出村庄，追赶象群去了。

A short while later, Mengzi gets back onto his feet and monkeys around in the yard. When he runs into the fence, his trunk gets stuck. Hoping to free his trunk, he pulls out the fence by accident. It is such a pain to get the fence off his trunk. Mengzi has to try many, many times.

He trots out of the village to catch up with his herd.

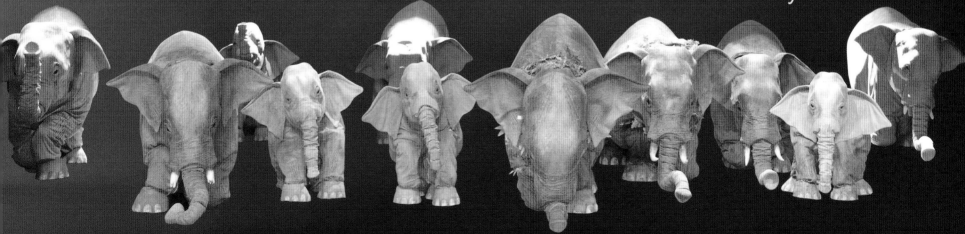

第五幕
Scene 5

请接受
"短鼻家族"
的敬礼

Please Receive the Salute of the "Short-trunk Family"

北迁的"短鼻家族"继续缓缓地走着。由于过度劳累，象妈妈迟迟没有奶水。两天没吃上奶的象宝宝，身体非常虚弱，已经走不动了。

The northbound "Short-trunk Family" continues its slow-paced journey. Due to exhaustion, Mama Elephant's milk runs dry. Baby Elephant, who has gone two days without nursing, has become critically weak.

象爸爸用鼻子推了推象宝宝，他挪动了一点，可还是无力地瘫在那里。象宝宝似乎真的不行了。

Papa Elephant goads the baby with his trunk. Baby Elephant budges a little but remains sickly on the ground. It seems that he will not hold out for long.

43

象妈妈不断地用鼻子托着他，让他起身，可是象宝宝好像一点力气也没有。

Mama Elephant keeps supporting him with her trunk, so that he may get up. It seems Baby Elephant has no strength left at all.

44

象妈妈无能为力，望着继续前行的象群，只好默默跟了上去。但象妈妈终究不忍心抛下象宝宝，又回到了他的身边。其他大象们也回到了象宝宝的身边，抚慰着他。

There's nothing Mama Elephant can do. She walks away sadly. But she then returns to Baby Elephant's side, for she cannot bear to part with him. All the elephants encircle the baby, consoling him.

45

象宝宝蜷曲在那里，根本没法动弹，虚弱无力。他发出了哀号。

Curled into a ball, Baby Elephant lies still. Lacking in strength, he wails.

这时，夜空中出现了无人机。象妈妈惊恐地看着这个机器，驱赶着它，保护着象宝宝。

A drone appears in the darkened sky. Mama Elephant looks at it with horror. She tries to chase it away, fearing it may harm Baby Elephant.

无人机侦察到了象妈妈和象宝宝的情况，于是指挥部派投食队投放了一批食物。有了食物的补充，象妈妈又有了充足的奶水，可以哺育象宝宝了。

Having seen it all thanks to drones, the headquarters send out a food delivering convoy. With food that helps with milk production, Mama Elephant now has more than enough milk to feed Baby Elephant.

象宝宝吃了奶，终于站起来了。所有的大象都跟着兴奋地吼起来。

这时，无人机经过。大象们盯着无人机，象妈妈跪在地上，扬起"短鼻子"，向无人机敬礼。

象宝宝和其他大象也相继跪在地上，扬起"短鼻子"，向无人机敬礼。

With milk in his stomach, Baby Elephant finally gets back onto his feet. All the elephants trumpet with joy.

The drone passes. The elephants follow it with their eyes. Mama Elephant kneels and raises her trunk, as if saluting the machine.

Baby Elephant and other elephants follow. They too kneel and raise their trunk to salute the drone.

49

《请接受"短鼻家族"的敬礼》

Please Receive the Salute of the "Short-trunk Family"

作词：石学海
作曲：杨 青 石天戈

中速

前奏

虽然 我们 不会 说 话 但 心里 什么

都 知 道 请接受 短鼻 家族

的敬礼 让我们 找到了共同的 家 请 接受短鼻家族的敬礼 让我们

找到 了共同的家 请接受 短鼻 家族 的敬礼 让我们

找到 了共同的家 让我们 找到了 共同 的 家

第六幕
Scene 6

回家的感觉真好

It Feels So Great to Be Back Home

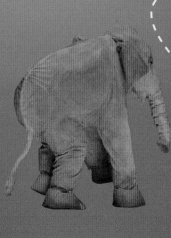

猛子在雨林中走着，拼命地追赶着"短鼻家族"。自从他离开象群后，才意识到"个体离不开群体"的"戒规"有多么重要。

　　Mengzi braves the rainforest, eager to catch up with the "Short-trunk Family." It is not until he left them that he realizes that an elephant is a social animal, and that social animals have rules to follow. He looks desperately for his herd.

象妈妈焦虑不安地来回巡视，她不希望失去任何一个孩子。她不停地呼唤着，仔细地听着。突然，远处似乎传来了微弱的声音。小象兰草也跟着象妈妈一起呼唤。

其他大象们也一起向远方呼唤了起来，仿佛嗅到了猛子的气息。

Mama Elephant's searching eyes dart about with distress. She does not wish to lose any child. Urgency bleeds into her calls, and she listens carefully for a response. All of a sudden, a weak noise seems to come from far away. Lancao joins her mother.

Others join as well. They direct their calls toward where the sound comes from, eager to detect Mengzi's scent already.

53

远处传来了猛子回应的声音。象妈妈接着又喊，大象们也跟着喊起来。呼唤声越来越近，越来越清晰。猛子回来了，大家都静了下来。猛子向前走了几步，一下子跪在象妈妈面前。

Mengzi's answering call reaches them from afar. Mama Elephant calls again, followed by others. The calls grow closer and clearer. Mengzi is back. Everyone falls silent. Mengzi walks toward Mama Elephant and falls onto his knees.

小象兰草冲上前去，蹲在猛子哥哥身旁。

Lancao rushes to her brother's side.

象妈妈拉起猛子，母子用头互相拱着，厮磨着，象群沸腾了。

回家的感觉真好！大象们跳起舞来。

Mama Elephant pulls Mengzi up. The mother and the son nuzzle and caress each other. The herd bursts with delight.

It feels so great to be back home! All the elephants break into dance.

《回家的感觉真好》
It Feels So Great to Be Back Home

作词：石学海
作曲：杨 青 石天戈

回家 感觉好轻松 睡觉睡到自然醒 回家

感觉好默契 不说话 意思全 懂 回家感觉好温暖

扫 码 听 歌

长辈 晚辈都平等 回家感觉好包容 有错

只是提个醒 回家感觉好平静 积

蓄能量再出征 回家的感觉真好 回家

的感觉真好 啊 回家的感觉 真好

《让全世界睁大眼睛》

Make the World Open up Its Eyes

作词：石学海
作曲：杨 青 石天戈

扫 码 听 歌

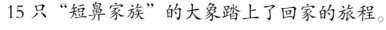

15 只 "短鼻家族" 的大象踏上了回家的旅程。

The fifteen elephants of the "Short-trunk Family" set off for home.

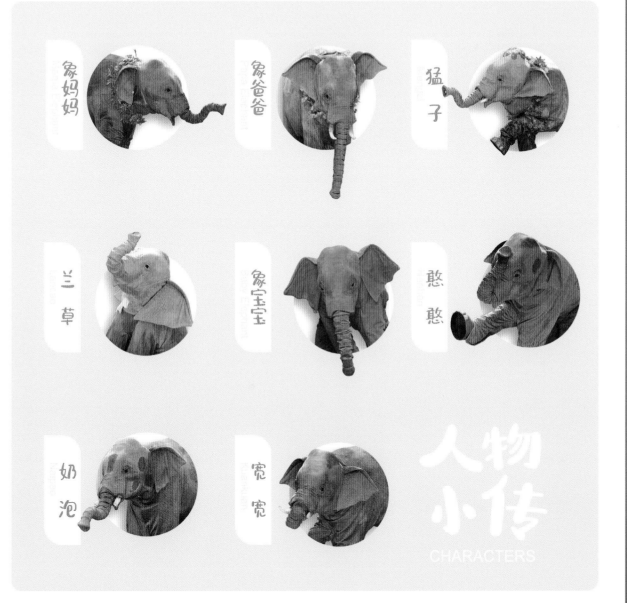

象妈妈
Mama Elephant

象爸爸
Papa Elephant

猛子

兰草

象宝宝
Baby Elephant

憨憨

奶泡

宽宽

人物小传
CHARACTERS

图书在版编目（CIP）数据

大象来了：汉英对照 / 中国野生动物保护协会，中国木偶艺术剧院编著 .
— 北京：中国海关出版社有限公司 , 2022.3
ISBN 978-7-5175-0569-3

Ⅰ . ①大… Ⅱ . ①中…②中… Ⅲ . ①长鼻目—普及读物—汉、英 Ⅳ . ① Q959.845-49

中国版本图书馆 CIP 数据核字 (2022) 第 027774 号

大象来了
DAXIANG LAILE

出 品 人：韩 钢
主　　编：武明录 韩 钢
策划编辑：史 娜 孙晓敏
责任编辑：刘白雪 景小卫
责任印制：赵 宇
摄　　影：武明录
绘图设计：邱 爽
英文译者：依 然
出版发行 中国海关出版社有限公司
社　　址：北京市朝阳区东四环南路甲 1 号　　　邮政编码：100023
网　　址：www.hgcbs.com.cn
编 辑 部：01065194242-7521（电话）
发 行 部：01065194221/4238/4246/5127/7543（电话）
社办书店：01065195616（电话）
　　　　　https://weidian.com/?userid=319526934（网址）
印　　刷：北京盛通印刷股份有限公司　　　经　　销：新华书店
开　　本：889mm×1194mm　1/12
印　　张：5　　　　　　　　　　　　　　字　　数：50 千字
版　　次：2022 年 3 月第 1 版
印　　次：2022 年 3 月第 1 次印刷
书　　号：ISBN 978-7-5175-0569-3
定　　价：68.00 元